Know
About Alcohol

Know About Alcohol

by Margaret O. Hyde

Illustrated by Bill Morrison

McGraw-Hill Book Company

New York • St. Louis • San Francisco • Auckland
Bogotá • Düsseldorf • Johannesburg • London
Madrid • Mexico • Montreal • New Delhi • Panama
Paris • São Paulo • Singapore • Sydney
Tokyo • Toronto

Library of Congress Cataloging In Publication Data.

Hyde, Margaret Oldroyd, date.
  Know about alcohol.

j362.2
H
cop3

  Bibliography: p.
  Includes index.
  SUMMARY: Discusses the effects of alcoholic beverages on the body, the
difference between use and abuse, and situations in which one must decide
whether or not to drink.
    1. Alcoholism-Juvenile literature. 2. Alcohol-Physiological effect-Juvenile lit-
erature. 3. Alcohol and youth-Juvenile literature. (1. Alcoholism. 2. Alcohol)
I. Morrison, Bill, 1935-   II. Title.
HV5066.H9 362.2'92 78-7988
ISBN 0-07-031621-X

123456789 BPBP 78321098

To Jared Adam Larrow

Contents

Foreword

Most of us learn about alcohol by watching other people drink. When the time comes for us to decide, we should decide for ourselves. The more we know, the better our decision. This book has the facts to help a young person decide. It also has the kind of attitude that can help our country have less people misuse alcohol.

> Morris E. Chafetz, M.D.
> Former Director of the National
>   Institute for Alcohol Abuse and
>   Alcoholism,
> President of the Health Education
>   Foundation

1.

Know About Alcohol

Have you ever seen a fisherman with a six-pack of beer on television? Or have you watched other television ads where beer is shown at ball games, on camping trips, at parties, at picnics, and many other places? Perhaps you have seen people drink wine at dinner or have cocktails before dinner. Beer is a beverage alcohol. Wine is another beverage alcohol. And the cocktails are made with different kinds of beverage alcohol. What do *you* know about alcohol?

What you know about alcohol may make it seem good, or it may make it seem bad. It is the way people drink it that is good or bad.

Drinking is a pleasant part of the lives of about seven of every ten adults in the United States. About one of every ten adults drinks more than

11

he or she should, however. The problem drinker, aged seven to seventy, gives alcohol a bad name. Alcohol gives the problem drinker an unhappy life. Alcohol controls his or her life.

You will have to decide at some time during your life whether you choose to drink. If you know the facts about alcohol, you can reach a more intelligent decision. If you choose not to drink, you can be more comfortable with that decision after you have examined the reasons why. And if you choose to drink, you can do so without getting drunk.

Beverage alcohol affects the lives of everyone. It even affects those who do not drink. The people who do not drink drive on the same roads as people who drink too much to drive safely.

Drinking means something pleasant for most people, but being drunk is frowned upon. This is especially true where wine or beer is sipped regularly with meals. Few people who have learned to drink this way have become problem drinkers if they do not drink between meals. Many children grow up to drink much the same way their parents do. Very young children learn about alcohol without knowing it. They watch their families drink, or listen to them talk about alcoholic beverages, even if they do not drink.

Some children join their parents in toasts at weddings and other celebrations.

Alcohol may be familiar to you in a form other than beverage alcohol. Actually, there is a whole chemical family of alcohols. Some kinds are being used in fuel experiments as a replacement for gasoline in automobiles. Rubbing alcohol has been used for many, many years. It helps to lower fevers and stop the ache in tired muscles when it is rubbed on the body. Many boys and girls use this kind of alcohol on their faces when they are troubled with acne. Some girls use it on their earlobes after they have been pierced for earrings. For young and old alike, rubbing alcohol is such a poisonous drink that it could kill a person who drank it.

Beverage alcohol, called ethyl alcohol, is the most common kind. It is usually the one people mean when they talk about alcohol. Strangely enough, many heavy drinkers know less about alcohol than those who just drink occasionally.

2.

Alcohol in the News

Drinking alcoholic beverages is not new. People were probably drinking them long before they knew how to make any records of the kind we call history. But alcohol is appearing a lot in the news today for many reasons. One reason is its increasing popularity among people of all ages. Another reason is the increasing number of accidents that are caused by people who have had too much to drink. Half of the people killed in this kind of accident are under the age of twenty.

You may have heard a newscaster say that more young people are drinking than ever before and they are drinking at an earlier age. Many people agree with this. They are concerned mainly about the young people who drink too much, and lose control. They worry also about a possible increase in the number of alcoholics or problem drinkers in tomorrow's world.

Children and alcohol are often in the news today. While articles do not give information to identify young people, they do mention cases such as the following as fairly common.

Tom is a ten-year-old who emptied the milk from his Thermos and filled it with liquor before he left for school each morning. By the time he reached school, he was feeling the effects of the sips he took on the school bus. His teachers noticed that he was sleepy in class but they did not think of alcohol as the cause of the problem.

After the bus driver reported Tom, the principal interviewed his parents. He discovered that the father was an alcoholic who drank so much that he never missed the liquor which Tom was using to fill his Thermos. Tom's mother worked and left before the bus arrived. Family therapy was suggested and an appointment was made at the local mental-health clinic.

According to some news reports, pre-teen and teenage drinking is such a serious problem that it should be considered a national crisis. Not everyone agrees with this.

A survey at high-school level revealed that most students had experimented with some form of alcohol. The PTA is conducting a program of alcohol education which stresses that alcohol use and abuse is a family affair.

Headlines of newspapers and magazines around the world reveal concern over the amount of drinking in other countries. For example, one finds "GERMAN CONCERN FOR CHILDREN MOUNTS" as the headline of an article which tells that a survey among 2,360 adolescents aged ten to eighteen years was recently conducted at a West German university. This survey showed that some teenagers spent up to $80 per month for alcoholic beverages. It also indicated that teenage alcohol abuse is most prevalent in the age group fourteen to sixteen, although even some eleven- and twelve-year-olds can be classed as alcoholics.

Some news articles about alcohol attempt to prevent future problems by educating young people. They warn against using scare tactics in an attempt to frighten people away from drinking. For many years, learning about alcohol meant learning all the bad things about it. A more meaningful approach is to help people drink sensibly, if they choose to drink.

New ways of thinking about alcohol problems are making the news, too. Some television and radio programming is helping to heighten the awareness of the American people about the growing problem of alcoholism. It points out that alcohol abuse is an educational problem which begins with an examination of one's own attitude toward life, one's self-image, and one's experiences, and the effect of alcohol on one's health.

While educational programming helps to present real-life attitudes about drinking, entertainment continues to show a drunk as funny, brave, or in some other false role. Would you make fun of a drunken person if you realized the person might be drunk because of an illness?

You can find many instances of alcohol in the news if you watch for them. Some are negative and some are positive. Some may be colored by emotional thinking. Watch for cases of this and for cases of truthful reporting about this subject that causes much disagreement.

Most experts agree that learning how to drink properly is important for those who choose to drink at all. They suggest that the time to prevent future alcoholism is the time before young people consider drinking. It is then that

they can read about alcohol with less pressure from those who want them to drink or not to drink.

3.

Alcohol in Bottles

Alcoholic beverages vary greatly in taste, use, cost, and the amount of alcohol they contain. Beer, wine, and liquor do not taste the same but they are all produced by a process known as fermentation. Fruit sugars, honey, sprouted seeds, or grain can be changed in the presence of small plants known as yeast to form beverage alcohol. You cannot see yeast plants, because they are so small, but they are always present in the air. The first wine was probably made when some sugar in fruit that had begun to age was changed in the presence of yeast plants. Today, wine making is a delicate art. Wine is usually made from grapes that are grown under special conditions to produce qualities that are wanted in wine.

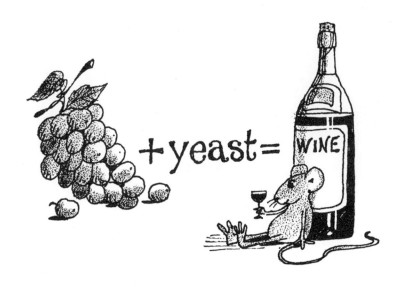

The fermentation process by which beer is produced is known as brewing. Grain cereals such as wheat, barley, and rye are combined with yeast under the direction of a brew master. This person follows a very special process often known only to him. Several thousand different varieties of beer are produced in countries in many parts of the world. They vary in ingredients and taste, but most contain from 3 to 6 percent alcohol.

If you compare beer to wine for alcohol content, a twelve-ounce bottle of beer contains about the same amount of alcohol as a five-ounce bottle of table wine. Stronger wines, such as sherry and port, are known as fortified wines. While most wines are made from grapes, sweet wines made from fruits such as apples and plums have become popular in recent years.

Brandies are made from fermented fruit juices, such as grapes, plums, and cherries. Sugar and flavorings are used with brandy to make liqueurs, sweet after-dinner drinks that are usually sipped in small quantities. Liqueurs may contain flowers, fruits, and/or herbs along with brandy. Sometimes they are made with other flavorings added to alcohol and water. Most liqueurs contain between 20 to 65 percent alcohol.

Since liqueurs are served in very small amounts and sipped slowly, the alcohol seldom has any effect. There are other sweet drinks which have become especially popular with young people. These drinks are made by combining liquor with a non-dairy milk. Often they are slightly stronger than wine in alcohol content. Such drinks appeal to people who have not learned to like the taste of other alcoholic beverages, and they have been criticized for their influence on young people learning to drink.

The taste of liquor such as scotch, gin, bourbon, and rye may not be pleasing to the majority

of young people when they are first introduced to them. The taste is cultivated for a variety of reasons, however. One of these is their effect on feelings; another is the wish to appear adult.

Liqueurs and liquor are produced by distillation. In this process, a fermented liquid made from starch such as grains is heated until it becomes a gas. Then the gas is collected and cooled, and it becomes a liquid of more concentrated form. This makes alcohol stronger.

People usually mix liquor, such as vodka, gin, whiskey, and rum, with water, soda, or tonic. If one compares beer, wine, and distilled liquor, the addition of mixers must be considered. Beer is 3 to 6 percent alcohol; wine is 7 to 20 percent alcohol; and liquor is 40 to 50 percent alcohol. A twelve-ounce bottle of beer, a five-ounce glass of wine, and one and one-half ounces of liquor are about the same strength. Since the liquor is mixed with water, tonic, fruit juice, or another liquid, the amount of alcohol in all these drinks is considered equal.

Many people find it surprising that there is as much alcohol in a bottle of beer as in a drink of liquor plus water. They drink beer in quantity and remark, "It's only beer." This expression becomes questionable when one knows that one beer, one glass of wine, or one gin and tonic may have equal amounts of alcohol.

The alcohol in all alcoholic beverages is the same kind. The other ingredients may have some effect on the digestive system, but the part that has an effect on the nervous system is the alcohol. Mixing drinks will not decrease or increase the effect, but there are many things that will. These are discussed in chapter five, on "For Those Who Choose to Drink: Learning Safe Drinking Habits."

4.

Alcohol in People

What happens when people drink alcohol as a beverage? This depends on many things, including the fact that each person is different, but the path which alcohol takes in the body is much the same for everyone.

Alcohol does not have to be digested, so it is absorbed quickly. About 20 percent of what is drunk is absorbed from the stomach and goes into the bloodstream. The rest of the alcohol goes into the small intestines and enters the bloodstream from there. Blood carries alcohol to all parts of the body, including the brain. Alcohol reaches the brain within minutes after it is drunk, and this is where the change in feelings takes place. At the same time, some alcohol which is passing through the liver is being changed from alcohol into water, carbon dioxide, and energy. The liver changes small amounts of alcohol each time the bloodstream carries its load through it. The part of the alcohol that is not changed continues to travel through the brain and other organs as the blood circulates through the body.

The liver works at a constant rate, so when there is more alcohol than the liver can change, the brain is exposed to alcohol again and again. The liver of a 150-pound man can change about 7 grams or ¾ of an ounce of alcohol per hour. For someone who weighs less, less alcohol is processed per hour. For example, the body of a young person weighing 100 pounds would process less alcohol per hour than that of a 150-pound man or woman. Since how one feels and reacts during and after drinking depends on how much alcohol reaches the brain, the size of a person plays an important part.

Some people note a change in mood after the first few sips of alcohol. If someone expects to feel relaxed or "high" when he or she drinks, this feeling may result even before any alcohol reaches the brain. Experiments in which people have been told that they are drinking alcohol

when they really were not have shown this to be true.

Sipping one drink, such as a twelve-ounce can of beer, has a mild tranquilizing effect on most people. Alcohol is actually a depressant, but it seems to act as a stimulant for some people when they first begin drinking. This happens because alcohol's first effects are on the part of the brain which controls behavior, such as self-control. With increasing amounts of alcohol in the bloodstream, vision becomes impaired, muscle coordination and balance are temporarily affected, memory decreases, and the mind has trouble getting things together. At this stage, one is considered drunk, which is a state of being physically and mentally handicapped.

Fortunately, most people do not drink enough at any one time to have the above symptoms. Unfortunately, drunks may also suffer from drowsiness or aggressive feelings or severe depression. Very heavy drinking can depress the deepest levels of the brain, causing coma or death.

Most people drink just enough to increase awareness and pleasure. They drink to feel more relaxed, more sociable, and to feel good. How much one drinks, how, when, and why all play an important part in whether drinking is an enjoyable experience or a sad one. If a person's mood is one of depression before drinking, alcohol may or may not increase that mood. No one knows beforehand how one will react to alcohol. Body rhythms, the time of day, attitudes toward drinking, drinking experience, and body chemistry are some of the things which play a part in individual reactions. So you can see that what happens when one person drinks the same amount of alcohol as another person may be quite different even when the kind of drink is the same. Suppose Jerry and Ken both drink the same amount of beer on the same day. Jerry is relaxing with friends and spends two hours drinking three cans of beer. Ken joins him after moving some heavy furniture. He is so thirsty that he gulps down his three cans of beer. You can easily see why Jerry and Ken have different effects from the same amount of beer.

How fast one drinks affects how much alcohol passes through the brain. Since the liver works at a steady rate, alcohol which is gulped has a different effect on the brain than when it is sipped. Some people describe fast drinking as jolting the brain with sudden gushes of alcohol. The speed of drinking is just one of the things

which make people react differently to the same amount of alcohol. In the above situation, the fact that Ken was tired added to the way his body reacted to the alcohol. He may have felt drunk, fallen asleep, been angry, started a fight, or acted in any of serveral different ways. Many of these reactions are well recognized by people of all ages as the state of being drunk even though they may have never tasted any alcohol.

Being drunk is being out of control of yourself. This is what most people want to avoid when they drink. This is what takes the pleasure out of drinking, and there are many ways to avoid this reaction.

5.

For Those Who Choose to Drink:
Learning Safe Drinking Habits

People who drink socially usually learn a few guidelines which help them to stay in control. Many young people learn unconsciously as they grow up because they notice what their parents do. Others learn from friends or older brothers and sisters. Some young people want to resist the pressure of friends who try to get them to drink until they are drunk, but they do not know how. The following guidelines help people to enjoy alcoholic beverages without getting drunk, and may help you and your older brothers and sisters.

Drink at meals or while eating foods such as cheese, Italian or French bread, and dips. Salty foods, such as pretzels and potato chips, in-

crease thirst so they are less desirable than other foods. Eating fifteen minutes before drinking helps more than eating while drinking, but there is no foolproof way to coat the stomach with milk or other substances so that one will not feel the effects of alcohol.

Avoid carbonated mixers. Carbon-dioxide bubbles increase the speed at which liquids travel through the stomach wall.

Realize that alcohol is a drug and do not drink after taking other medication. Alcohol is a depressant drug. If someone has taken a tranquilizer or other medicine to depress body activities, that drug plus alcohol may add up to more than the sum of the two. Alcohol in combination with stimulant drugs can cause a false sense of security and other, more serious problems.

Responsible drinkers measure the amount of alcohol when making a drink rather than pour it directly from the bottle. If a person is not trying to get drunk, the appearance of the drink can satisfy even when the alcoholic content is low. Using a small-diameter glass and plenty of ice helps to make such a drink.

Experienced and careful drinkers know that beer, wine, and liquor can all produce drunkenness and that an amount which does not affect a person on one day may have a different effect on the same person on another day.

Small people feel the effect of alcohol faster than large people. The bodies of large people contain more water than those of small people, so alcohol is diluted more when they drink. This is not true of fat people, since fat does not dilute alcohol.

Social drinkers make drinking part of another activity such as playing a game, talking with

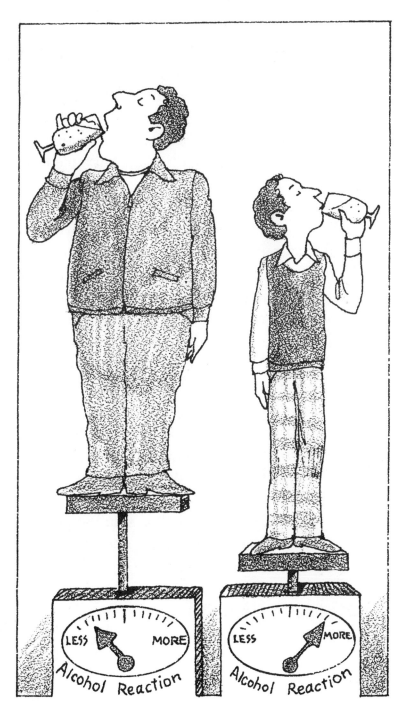

friends, eating, or planning an event. Drinking for relaxation is more likely to take place in a quiet setting than in one where loud noises or loud music may increase one's sense of isolation. Soft music diminishes tension, making the reaction to a small amount of alcohol a pleasant experience.

Tolerance plays a part in learning to drink safely. As mentioned earlier, people have individual differences in the amount of alcohol which affects them. While one kind of tolerance is acquired, there is a kind of "built-in" or natural tolerance which is present even when one first begins to drink. In general, an inexperienced drinker's body would show the effects of the alcohol longer than an experienced one. Drinking produces tolerance for which an increased amount of alcohol is required to produce the same effect. Young people would normally lack this kind because of their inexperience with drinking.

Since feeling relaxed with a small amount of alcohol is more desirable than needing a large amount to produce the same effect, tolerance is not necessarily a good thing. In fact, some experts feel that a high tolerance for alcohol is really a first symptom of approaching alcoholism.

The importance of tolerance should be considered in responsible drinking, especially when driving is involved. The amount of alcohol which causes problems for one person may be quite different from the amount for another.

Learning to drink wisely means learning to drink without getting drunk. Tolerance makes the definition of being drunk confusing. Consider someone's Uncle Joe, who is said to be able to "hold his liquor." Uncle Joe drinks heavily, but he never appears to be drunk. While drunkenness is defined as temporary loss of control over physical and mental powers due to the drinking of too much alcohol, Uncle Joe appears to have control. Therefore, he is not drunk by this definition. By legal definitions, drunkenness is defined by the amount of alcohol in the blood. Uncle Joe may appear to hold his liquor well, but he may not pass the tests of coordination and judgment needed to drive safely. This may make him especially dangerous to society. Rather than considering this kind of drinker a hero, most young people recognize him as a problem drinker. That is, he is dependent on alcohol, or has built up so much tolerance that he must drink more than the normal amount before he reacts.

Millions of Americans who choose to drink do so in a responsible manner. They do this not only because they have learned of the dangers of alcohol and alcoholism. They know also that anyone who chooses to drink has a responsibility not to harm his or her own self or other people. If you choose to drink when you reach the age at which you would normally decide, remember the positive and the negative things.

6.

If You Choose Not to Drink

Whether or not one chooses to drink is a private decision. Your choice should not be questioned. But this is not always the case. Many people who choose not to drink find it helpful to explore their reasons for themselves and feel secure with them. Then they feel comfortable about just saying, "No thank you."

Reasons for choosing not to drink are numerous. Some dislike the taste or find the effects of alcohol are unpleasant. Many people have strong religious convictions which forbid the drinking of alcoholic beverages. Someone who is close to an alcoholic may reject drinking because of the trouble that it has caused the alcoholic and that person's family. Fear of breaking the law or moral principles about breaking

41

the law are reasons some young people abstain.

Individuals who are non-conformists may take pride in their ability to refuse to go along with a crowd of drinkers. They are proud of coping with life without the use of chemical crutches. However, many adolescents find it difficult to say no to a group of friends for fear of being different. Many young people who do not really want to drink think they must go along with the crowd. Actually, many studies show that those who choose not to drink remain popular with their friends, although it is true that these people tend to associate with other non-drinkers more than with drinkers.

Consider the case of Sue, who is being pressured to drink at a party. Some of the boys who have a drinking problem want her to get drunk along with them. Sue is afraid they may get her to say yes to something she would refuse if she stayed sober. And since Sue has never approved of drinking in the first place, she wonders why she is even considering impressing these boys. As far as she is concerned, she is out with the wrong people.

Diane has firm convictions about drinking, but she is as wrong as those who pressure a non-drinker to join them. Diane tries to reform any-

43

one who drinks in front of her, even when the person is just sipping a beer. She expects her friends to respect her right not to drink but she does not respect their right to choose what they feel is right.

Many young people are refusing to drink today because they are aware of the problems of drinking and driving. While various groups campaign against pollution and nuclear power plants, others are campaigning against the drunken driver. About 50 young Americans will be killed today on the highways and about 250 will be maimed or disfigured. About half of these people will be killed or injured because someone was driving when drunk. Some of the young people who campaign against this tragic situation refuse to drink at all.

Young people who are concerned about the drinking habits of their parents provided the idea for a program that was carried out by an organization known as Parents Without Partners. The organization's International Youth Council asked their parents to become involved with the problem of drinking too much. One approach was especially successful. A group of Parents Without Partners in Pueblo, California, acted upon a remark of a man who had once had

a drinking problem. He asked if there were any parties where alcohol was not served. This group of Parents Without Partners held some and found that everyone could have a good time without alcohol.

For people who choose not to drink, it may seem that they are alone in their decision. Actually, about one-third of the people in the United States are non-drinkers. Respecting the non-drinkers in any group is a greater sign of maturity for those who choose to drink than trying to persuade the others to change their decision.

The use of alcohol does not make a person sexier, more glamorous, more adult, or more mature. You will not have to worry about how much is too much if you choose not to drink. And friends who are worth having will respect your decision not to drink.

7.

Alcoholics: Young and Old

Do you know an alcoholic? Many young people are concerned about someone they know who may have a drinking problem. That someone may be young or old.

Older alcoholics have been identified by some of their common symptoms. Many suffer from liver disease. This organ is involved in breaking down alcohol so that it can be discarded by the body. After many years of too much work, the liver grows fatty and may be enlarged. Some of the liver cells die and scar tissue replaces the active, living cells that function to keep the body healthy.

Too much alcohol over a period of years takes its toll on many health fronts. One of these is the pancreas, an organ that is important for the digestive system. Even people who drink too much only occasionally may notice that they have difficulty digesting fatty food the next day. Alcoholics frequently suffer from damage to the pancreas.

In severe alcoholism, there may be brain damage, too. There are billions of cells in the brain, so a few drinks that kill thousands of them seem unimportant. But in the case of an alcoholic who has been drinking a large amount for a long time, brain damage may be so bad that it is obvious to anyone who knows him or her. Some people who were very bright cannot work at jobs that require little skill after suffering such brain damage.

For young people, these and other effects of many years of heavy drinking seem far away. Young alcoholics do not usually suffer physical damage other than that brought about by poor nutrition. But one recognizes them in other ways.

Alcoholics are people who drink excessively and whose dependence on alcohol interferes with their physical or mental health. Alcoholic people do not always seem alcoholic, for most of them try very hard to hide their illness. There is no simple test which fits each one, but there are some signs which point to a serious drinking problem. First, when most alcoholics drink, they lose control of drinking and continue until becoming drunk or even unconscious.

Young alcoholics are recognized by their disturbed family or social relationships, alcohol-related car crashes, drunkenness, delinquency, or anti-social behavior. One or two instances of drunkenness do not mean that a person is an alcoholic, but this is not social drinking either. Pro-alcoholic young people rarely admit to their problems, but excessive use is often apparent.

Even the experts do not agree on how to define an alcoholic, and it may be there is no definition to fit everyone. An official of the Women's Christian Temperance Union has been quoted as saying that an alcoholic is anyone who drinks alcohol. This was in the year 1974. Few people agree with this definition, but many experts feel that answers of "Yes" to three or more of the following questions may indicate alcoholism:

YES NO

____ ____ 1. Does the person talk about drinking often?

____ ____ 2. Is the amount of drinking increasing?

__	__	3.	Does the person sometimes gulp drinks?
__	__	4.	Is a drink often used as a way to relax?
__	__	5.	Does the person drink when alone?
__	__	6.	Are there memory blanks after drinking?
__	__	7.	Does the person need to drink to have fun?
__	__	8.	Have hidden bottles been found?
__	__	9.	Does the person drink in the morning to relieve a hang-over?
__	__	10.	Does the person miss school or work because of drinking?

You cannot tell with any certainty whether or not a parent or friend is an alcoholic by this test, but it may help you to find out whether or not one has a serious problem. Sometimes people imagine that others drink too much and worry without reason. Sometimes, if there is reason to be concerned, a person can find help by discussing the problem with a counselor, a hotline worker, or social worker. Often a family doctor can be helpful.

There is much controversy about the number

of young alcoholics. Almost everyone agrees that young problem drinkers are increasing in number. The fact that there are more young people in the world today and more of them who communicate their problems to counselors, hot-lines, and other helping agencies makes some experts question whether the problem drinkers are more numerous or whether people are just more aware of them.

Many surveys have been made to try to find out more about drinking among young people in an effort to prevent alcoholism. The surveys are usually made in schools so that those who have dropped out are not covered. Since the extent of problem drinking is thought to be high among those who drop out of school, the amount of problem drinking may well be higher than is shown in the surveys. Even so, most of the studies indicate that the amount of problem drinking at the high-school level is very high. This is one of the reasons that many efforts are being made to reach people as young as elementary-school age with information about the use and abuse of alcohol. *Hardly anyone plans to become a problem drinker. Knowing how to avoid this comes before one begins to drink.*

Yes, there are some ten-year-olds who abuse alcohol. But most people who decide to drink do so for social reasons at a much later age. One in ten who drinks eventually becomes an alcoholic. No one knows exactly why. Some individuals have a particular kind of body chemistry or have emotional problems of a kind that makes them more likely to become alcoholics than others. Children of alcoholics have a 50 percent greater chance of becoming alcoholics than others. While research is still being done to try to determine if alcoholism has something to do with heredity, those who know about alcoholism in their families are being warned to be very careful about their drinking patterns.

Many studies have been made about alcoholism and how people feel about drinking. In countries where people drink wine only with meals, as in Italy, alcohol is considered just part of eating and of being sociable. Children who grow up in this kind of setting, where getting drunk is frowned upon, are very unlikely to become alcoholics. Orthodox Jews and Chinese who cling to old customs seldom have problems with alcohol, and here, too, heavy drinking is not acceptable behavior. When the children and grandchildren of these ethnic groups adopt

American customs in which heavy drinking for the sake of drinking may be popular, Italians, Jews, and Chinese have problems with alcohol.

Whether alcohol is a disease or a learned behavior is a point of argument even among the experts. Sometimes a person may drink heavily beginning with the first taste, so this cannot be considered a learned behavior. People who are often in situations where there is a great deal of drinking tend to have more problems with their drinking behavior than others, and this could be learned behavior. Whether or not alcoholism is called an illness or a learned behavior, the image of an alcoholic as a weak-willed person or a bum is changing. A drunk is no longer funny.

For the ten million people who have already become alcoholics there are many types of treatment, including Alcoholics Anonymous, whose address is listed at the end of this book. Local chapters are listed in telephone books and meetings are open to all alcoholics who seek their help. As the name implies, the person can remain anonymous, since only first names are used. In some areas, young alcoholics have found help by attending meetings and in some parts of the country there are special groups for young people. In Los Angeles, California, there

are at least four chapters especially for young people who are willing to accept the program.

The treatment of alcoholism in young people is difficult and often unsuccessful. Alcohol abuse may be just the symptom of a hidden problem. Mental-health workers and doctors who specialize in emotional problems may be able to help *if* the person wants help. Very often, a person drops out of a treatment program rather quickly.

Groups of young people who have problems with alcohol are sometimes helped by an understanding counselor or social worker and by each other. Teenage counselors, or peer counselors, have been more successful in many cases than older people.

In some programs, young people are being trained to talk to those of elementary-school age in an effort to prevent alcohol abuse. Young counselors help them to understand that using alcohol as an escape from problems only postpones the time when they will have to work these problems out and at the same time have to deal with an alcohol problem, too.

The number of children below the age of twelve who have already become involved in alcohol problems is unknown. Dr. Morris E.

Chafetz, former director of the National Institute on Alcohol Abuse and Alcoholism is concerned that the number of cases of alcoholism in children between the ages of nine and twelve may have increased tenfold in recent years. Even though the actual number of cases of alcohol abuse in this age group may be small compared with the number of children of this age, it is still a serious problem.

Since there are an estimated 500,000 young people who have become problem drinkers before they have reached the age when drinking is legal, alcohol deserves its title as the number one drug problem among the young.

Many adults are becoming more and more aware that they need to act as models for their children, especially since studies show that most children follow the drinking patterns of their parents. The exception to this is the number of alcoholics who come from homes where there is no drinking at all. Studies show that most alcoholics come from homes either where alcohol is forbidden or where one or both parents are alcoholics. These studies also indicate that parental drinking patterns are the most important factor in determining the kind of drinking children will do.

Here is the actual story of the son of an alcoholic who began drinking at the age of twelve. It shows how a young person became an alcoholic.

I'm Charles, 17, a high-school student and an alcoholic. I started drinking at the age of 12. I was hanging around with an older group of kids and I drank to be 'cool.' Nothing much happened from my drinking until I was 14. By that time I looked older than I was, and I could buy beer or liquor without any trouble. The summer before I entered the eighth grade I started working in a gas station in a neighborhood where the people are very alcohol-oriented. My boss drank a lot and he always had beer around the station. He encouraged me to drink. . . . I changed a great deal that summer and took a turn toward alcohol. I changed my attitude towards other people, got sloppy and careless, and started getting into fights.

By the time I started school that fall, I was drinking so much that I needed something every morning for some get-up-and-go. I started pitching in with other boys in the eighth grade to buy a case of beer so we could drink during school. We would hide it in some bushes near the school and go up there during recess. By lunch we would be pretty well totaled. The group of us started growing in number. The playground supervisor began wondering why half of the eighth grade were wandering up the hill during recess. Finally, he took a look—and found the beer. He let us off easy, just told us not to do it anymore. When we asked him what to do with the leftover beer, he said to put it in his car. . . . It was a fun year. I got drunk every weekend. . . . It seemed like everybody in the eighth grade graduating class was drunk. One boy fell down as he was about to get his diploma.

I left home that summer, caddied for a country club and lived in the clubhouse. At night, friends and I would go bar hopping. One night in a Washington bar, I got up to talk to a pretty girl who had smiled at me. The next thing I knew, I was coming to in an alley. . . .

I moved back with my father and entered junior high school. I started smoking pot, but it was making me lose the desire to do things. So I cut it out and started drinking booze again. . . . By the time I got to the tenth grade, I was keeping a 6-pack of beer in my locker at all times. . . . I went out for the football team and would have been a starter if I hadn't been drinking. . . .

I turned into a super-derelict. I didn't care about school or anything, just drinking. I started getting the DTs. I stayed drunk all the way through the next summer. In the fall, I went out for football again. The coach kept me on the bench most of the time because he knew I was drunk. He put me in one game at the very end as a fullback. I couldn't get going and about 20 guys piled on me. . . .

It took an accident to get me straightened out. I was driving along and started to hallucinate and crashed into the median strip of a highway at about 50 miles an hour. I was knocked unconscious for about 10 minutes. When I came to, I decided it was time to do something about my drinking. Since I already belonged to Alateen, I knew about Alcoholics Anonymous and that's where I went for help. Now I haven't had a drink for 6 and a half months and I'm staying sober one day at a time.*

Alateen is an outgrowth of Al-Anon, the part of Alcoholics Anonymous that was organized for husbands and wives of alcoholics. Alateen helps teenage sons and daughters cope with their problems and meet others who have similar

*Reprinted courtesy of *Alcohol Health & Research World*, Summer 1975.

problems. Local groups can be found through listings in telephone books or by contacting Al-Anon Family Group Headquarters listed on page 75. While people under twelve cannot attend Alateen meetings, they can write for booklets that may be helpful. (See page 75 and 76.)

Teenagers in Alateen learn that an alcoholic is sick. He or she is in the grip of a terrible disease that cannot be controlled by the individual any more than a person can cure his or her own tuberculosis or diabetes.

What causes alcoholism is a question with many answers, and it may vary from one person to another. But if a family of an alcoholic is willing to learn about alcohol and acquire an attitude based on knowledge, the alcoholic's chances of recovery are much greater than if his or her family is ignorant about alcohol.

Whether or not a person who begins to drink will eventually become an alcoholic depends on many things, such as the amount of alcohol he or she drinks, the reasons for drinking, the way it is done, the time, and where it takes place. The *person* is especially important for reasons mentioned earlier. If one has a good self-image, knows how to make responsible decisions, has

self-confidence and factual information about alcoholism, the chances of becoming an alcoholic are lessened.

8.

How Would You Decide?

Suppose you find yourself in any of the following situations at some future time. What would you decide to do? Sometimes there is no right or wrong answer, but some answers are better than others.

You have been babysitting for some neighbors. When they return from their party, you can tell that the man is drunk. He says he will drive you home after he gets a cup of coffee. You know that the cup of coffee will not make him sober. What would you do?

WOULD YOU DECIDE TO:
1. Ask to use the phone and call your parents to take you home;

2. If the wife is sober, ask her to drive you home;
3. Call a friend or relative and ask to be driven or walked home;
4. Try not to hurt the man's feelings, even if you have to drive with him;
5. Arrange beforehand to stay overnight if you have reason to believe you might run into a problem, or

offer to sleep on the sofa if you have
not made previous arrangements.

DISCUSSION:
You should not allow your judgment to
be influenced because you might hurt
someone's feelings. In this case it might
be a question of safety for you and for
others on the highway. If the man is
drunk, he will not resent your objections
after he has had a night's sleep. One
babysitter remarked that her life was
worth more than the money she got for
sitting.

Your father is an alcoholic. This is creating
serious problems for him and for the entire family. You want to help him.

WOULD YOU DECIDE TO:
1. Tell him that he would stop drinking
   if he loved you;
2. Learn all you can about alcoholism
   through an organization such as Alateen so you can help yourself;
3. Beg him to go to an Alcoholics
   Anonymous meeting;

4. Hide the liquor;
5. Ask your mother to go to Al-Anon;
6. Learn as many facts as possible about alcoholism.

DISCUSSION:
1. Your father probably does love you. His alcoholism is an emotional problem and/or physical problem for which he needs help. A family's attitude can help him.
2. If you are over twelve, Alateen meetings can be helpful. If you are younger or there is no group in your neighborhood, contact Al-Anon Family Headquarters, P.O. Box 182, Madison Square Station, New York, New York 10010. Some alcoholics resent their family contacting such an organization. You may need the help of a counselor at school, a religious organization, a community center, or someone you can trust to talk with.
3. *Begging* a parent to go for help seldom works.
4. An alcoholic will find some other

sources when supplies are hidden. This may cause increased hostility.

5. If your mother is interested in finding help, Al-Anon may be an answer.
6. Contact the National Clearinghouse for Alcohol Information, P.O. Box 2345, Rockville, Maryland 20852, with a request for booklets for young people.

Your friend Mary has helped herself to a few beers from the refrigerator. She suggests you both ride your bicycles to the store for some potato chips. What would you do?

WOULD YOU DECIDE TO:
1. Go to the store alone;
2. Let Mary go to the store alone;
3. Find something else to eat;
4. Tell Mary you have to go home.

DISCUSSION:
If possible, persuade Mary to find something else to eat. If necessary, go home.

Your older sister is drinking secretly and she

wants you to join her. You know that she is troubled, but you cannot persuade her to ask for help. Last night, she told you about a wild drive with some of her friends who had been drinking. What would you do?

WOULD YOU DECIDE TO:
1. Continue to talk to your sister about her problems;
2. Drink with her so she will not pester you anymore;
3. Talk to someone your sister respects but who is not part of the family;
4. If you think your parents will understand, talk to them;
5. Tell your sister that her problem is so serious that either she must get help or you will have to talk to someone about it.

DISCUSSION:
While your decision depends somewhat on the individual situation, the fact that your sister is telling you about her problems may be a way of asking you to help. If you choose number 5, and carry

through if she does not get help herself, you will be doing what must be done to prevent further trouble.

You are waiting for the school bus when the older brother of a friend drives up to you. He offers to drive you home. You would rather ride in his car than wait for the school bus, but you can tell that he has been drinking heavily by the way he acts and talks. What would you do?

WOULD YOU DECIDE TO:
1. Tell the drunken driver you plan to meet a friend on the bus;
2. Tell the driver you do not want to ride with him because he has been drinking;
3. Go with the driver and hope for the best;
4. Tell the driver you have other plans.

DISCUSSION:
If you tell the driver you have other plans, you will be telling the truth and can then avoid riding with a drunken driver. Trying to reason with a person who is drunk seldom accomplishes any-

thing. Riding with a drunken driver is obviously dangerous.

You have reached the legal drinking age but you have not had much experience with alcohol. You have just finished a can of beer. Someone wants you to mix drinks but you think this might make you drunk.

WOULD YOU DECIDE TO:
1. Avoid mixing drinks;
2. At a later date, find out whether or not it is true that mixing drinks is dangerous;

3. Decide to limit the amount you
   drink.

DISCUSSION:
Limiting the amount of alcohol is more
important than whether or not one
mixes different kinds of beverages.

𝟗.

Alcohol: Good or Bad?

Now you know something about alcohol. And you know enough to be aware that the effects of alcohol differ for different people. They even differ for the same person at different times.

So far, no modern nation has been able to control the misuse of alcohol by prohibition laws, by taxing alcoholic beverages, by changing age restrictions on who may buy them, or by making rules about advertising them. Only people who know something about alcohol can make it a good thing or bad thing for themselves. Only people who know something about alcohol can know what is best for them.

Suggestions for Further Information

Inexpensive material is available from Alateen:
Al-Anon Family Group Headquarters, Inc.
P.O. Box 182
Madison Square Station
New York, New York 10010

Local services may be available from some of the following:
State Division of Alcoholism
State Department of Education
State Department of Mental Health
State Safety Council
Local Chapter of the National Council on Alcoholism
Local Chapter of the American National Red Cross
Local Chapter of the Jaycees

Contact Alcoholics Anonymous or the Alateen local chapter through your telephone book to learn about meetings.

Write for free and inexpensive booklets from:
The National Clearinghouse for Alcohol Information
tion
P.O. Box 2345
Rockville, Maryland 20852

Suggestions for Further Reading

For ages 6 to 9
Seixas, Judith S., *Alcohol: What It Is, What It Does,* New York: Greenwillow Books, 1977.

For ages 12 and up
Hornik, Edith L., *You and Your Alcoholic Parent,* New York: Public Affairs Pamphlet No. 506, 1974.
Hyde, Margaret O., *Alcohol: Drink or Drug?* New York: McGraw-Hill, 1974.
Lee, Essie, *Alcohol: Proof of What?* New York: Julian Messner, 1976.
Milgram, Gail Gleason, *What Is Alcohol? And Why Do People Drink?* New Brunswick, N.J.: Center for Alcohol Studies, Rutgers University, 1975.
Silverstein, Alvin, and Silverstein, Virginia B., *Alcoholism,* Philadelphia, Pa.: J. B. Lippincott Co., 1975.

INDEX